Production Cost Trends and Outlook:

A Study of International Business Experience

By James R. Basche, Jr.

*A Research Report from The Conference Board's
Division of International Affairs
G. Clark Thompson, Senior Vice President*

Contents

	Page
FOREWORD	v
1. INTRODUCTION	1
2. OVERALL PRODUCTION COST INCREASES	6
3. COMPONENTS OF RISING PRODUCTION COSTS	14
Raw Material Costs	15
Labor Costs	18
Energy Costs	22
Other Components of Rising Production Costs	23
Components of Rising Production Costs Outside the Home Country	24
4. EFFECTS OF HIGHER PRODUCTION COSTS	27
Reasons for No Changes	27
Changes Resulting from Higher Production Costs	30
Relocating Manufacturing	30
Technology Changes	31
Other Changes	32
5. PRODUCTION COSTS IN THE FUTURE	33
Expected Labor Cost Increases	35
Expected Raw Material and Energy Cost Increases	36
Possible Changes Resulting from Future Production Cost Increases	37
Effects on New Manufacturing Facilities	37
Other Effects	38
No Changes	39

Tables

1.	Five-year Production Cost Increases—Percentage	7
2.	Five-year Production Cost Increases—Percentage Range	9
3.	Five-year Production Cost Increases—Rubber Industry	10
4.	Five-year Production Cost Increases—Chemical Industry	10
5.	Overall Cost Relationships—1976	12

6.	Summary of Overall Cost Relationships	13
7.	Five-year Increases in Material Costs—Percentage	16
8.	Index of Unit Labor Costs in Manufacturing	18
9.	Manufacturing Labor Costs—1972-1975	19
10.	Five-year Wage Increases—Percentage	20
11.	Percentage Range, Five-year Wage Increases by Country	21
12.	1975 Indexes of Manufacturing Output and Unit Labor Costs	21
13.	Labor Costs Compared, 1975 or 1976 Using Home Country as 100	22
14.	Estimated Percentage Increases in Overall Production Cost over the Next Five Years	34
15.	Estimated Labor Costs per Unit of Output—a British Company	35

Charts

1.	1975 Unit Labor Costs and Production Index	2
2.	1975 Hourly Wage Rates and Production Index	3
3.	1975 Raw Material Prices and Production Index	4
4.	Six-year Production Cost Increases—a U.K. Consumer Products Company	9
5.	Raw Material Prices	17
6.	Five-year Percentage Increases—Costs and Output (U.S. Company)	25
7.	Comparison of Productivity and Labor Cost in 1975 (Italian Company)	26
8.	1976 and 1980 Wage Relationships for a Dutch Company	36

FOREWORD

PRODUCTION COSTS traditionally have been an important, and sometimes decisive, factor in determining where manufacturing firms locate their producing facilities. In recent years, production costs, fueled by higher energy costs, higher raw material costs, and higher labor costs, have risen rapidly in all countries of the world.

These rising production costs have led many chief executives to reexamine the locations selected for their new manufacturing units, to review cost factors at existing facilities to see where they might be more tightly controlled or even reduced, and, in a few cases, to move factories to lower cost areas or to terminate manufacturing at high-cost facilities.

Despite increasing production costs, few executives are prepared to move their facilities entirely out of countries where they are currently manufacturing. Costs continue important, but in many cases are no longer decisive. Tariff and nontariff barriers, local content requirements, local tastes and local regulations, as well as other factors, can all be equally or more important than production costs in determining the location of production facilities.

Chief executives from ten developed countries were asked to describe the recent experiences of their firms with rising production costs at home and abroad and to assess the effects of these increases on their companies' operations. This report is based on their descriptions and assessments. We are grateful to them for the assistance which made this study possible. The report was written by James R. Basche, Jr., Senior Research Associate in the Department of International Management Research under the direction of James Greene, Department Director, and the overall supervision of G. Clark Thompson, Senior Vice President for the Division of International Affairs.

KENNETH A. RANDALL
President

1.
Introduction

OVER the past four years, chief executives of international business firms have been confronted with rapidly rising costs in producing goods and services. The quadrupling of oil prices, a period of material shortages, unstable exchange rates, high inflation, and rapidly escalating labor costs have threatened the life functions of many firms—growth, profits and even survival.

These cataclysmic changes in the environment have been well-documented and represent subjects for ongoing debate. But what has been the experience of individual companies producing in many countries under such uncertain conditions? And how do the senior executives of these firms assess their condition and their options when production costs rise unevenly in their world operations?

Chief executives of ten countries were asked to describe the multinational experience of their billion-dollar firms with production cost rises, and to assess both the causes and consequences of these increases.

Production costs viewed within the framework of national geographic boundaries obviously do not change uniformly. The size of the existing pool of labor, both skilled and unskilled, for work in industry; the levels of skill, education and training of workers; the availability within the nation's borders of raw materials for general production and for energy; the level of technological development; and many other factors combine in different proportions to determine the overall level of production costs in each country. They also determine the cost variations among industries within a specific country and even among companies in a particular industry.

Chief executives report widely varying experience, but the overall figures amply demonstrate the conditions found by senior executives. At the end of 1976, the Organization for Economic Cooperation and Development (OECD) published figures showing that unit labor costs for a selected number of countries had risen substantially more than had production between 1970 and 1975 (Chart 1). Unit labor costs, of course, are affected by both wage rates and productivity rates. Hourly wage rates by themselves rose much higher than unit labor costs for the OECD countries, with the exception of Sweden, during the same five-year period. Executives in the United Kingdom, Italy and Belgium experienced a doubling of unit labor costs in five years. Chart 2 shows the hourly figures prepared by the OECD for ten countries.

The rapid rise in the cost of raw materials also outdistanced production rises in the first half of the 1970's for the countries on which the OECD reported at the end of 1976 (Chart 3). In the cases of Belgium and France the rise in hourly wage rates since 1970 (Chart 2) outpaced the rise in raw material prices while the reverse was true in Italy and the United Kingdom.

Despite the general rise in production costs around the world, many senior corporate executives believe other factors affecting their interna-

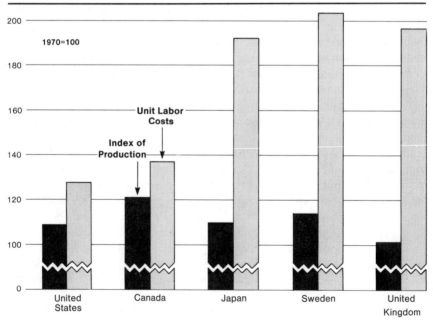

Chart 1: 1975 Unit Labor Costs and Production Index

Source: Organisation for Economic Co-operation and Development, Main Economic Indicators, December, 1976

Chart 2: 1975 Hourly Wage Rates and Production Index

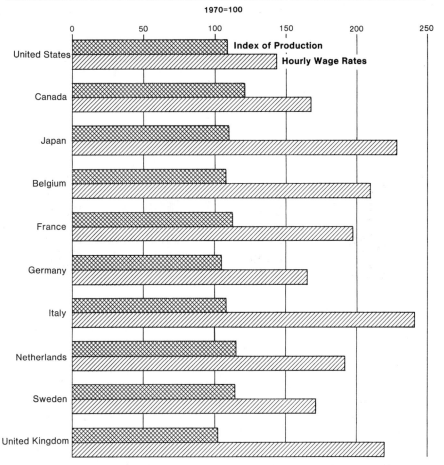

Note: Japanese wage rate is calculated on monthly basis.
Source: Organisation for Economic Co-operation and Development, Main Economic Indicators, December, 1976

tional business are more important. The comparison of production costs among manufacturing units in different countries is often given a low priority for their firms, and in some instances such intercountry comparisons are not made. Executives in these latter firms feel that regardless of high, and increasing, production costs in most countries, other factors affect competition, too. These factors may be so important that their firms must continue to produce in those countries if they are to continue to serve those markets.

Chart 3: 1975 Raw Material Prices and Production Index

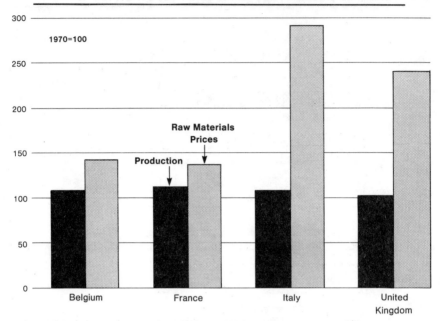

Source: Organisation for Economic Co-operation and Development, Main Economic Indicators, December, 1976

Among the factors reported as overriding production cost considerations are local government regulations requiring local production of some products, or at least the production locally of some product components. Import barriers have been erected by some governments to protect their high-cost producers. Local production is often required because some products, such as those made by the food and beverage industries, do not travel well. Furthermore, transportation costs for some companies have risen faster than production costs so that higher production costs locally may be less significant for the total price of a product than the cost of transportation.

Government regulations for testing or government specifications for a product, such as drugs or safety devices, may require local production, and these requirements outweigh production costs. Local tastes also make a difference. A beverage produced in one country may have little or no market in another; one product cannot be substituted for another. In such cases, local production for local tastes is more important than production costs which may be higher locally than in the company's home country or elsewhere. Thus, in many industries, local specifications or local tastes result in products that may be similar but not identical. Where this is true, many executives feel a comparison of local production costs among countries has limited value. Production costs are not the determining factor on how the market is to be served in such situations.

Because production costs are rising on a worldwide basis and because this rise in costs tends to affect all companies—domestic and foreign, competitors and noncompetitors—in essentially the same way, recent production cost rises, many businessmen believe, are less significant than they might be if the cost rises applied only to foreign companies, or to one company and not to its competitors.

To examine the effects of recent production cost increases on company operations and plans, The Conference Board solicited the experience of chief executive officers in ten countries with production cost changes in their home country and in operations outside their home country. Executives in eight European countries (Belgium, France, Germany, Italy, the Netherlands, Sweden, Switzerland, and the United Kingdom), Canada and the United States supplied the perspectives and data on which this report is based.

2.
Overall Production Cost Increases

THE RANGE of production cost increases confronting these executives in their own countries and abroad is wide. While a Swiss chemical company president could report a modest 15 percent increase over five years, the chairman of a British consumer products company cited a 217 percent increase over six years. Both executives were reporting these figures for their production facilities in Switzerland. The highest five-year increase was 213 percent, reported by an American general industry executive for his production facilities in Belgium. Figures reported by a number of executives are shown in Table 1 in detail and are summarized in Table 2.

For production units in the United States, reading across for example, 23 U.S., Canadian and European executives report cost increases ranging from 18 to 115 percent. Of the 23, fourteen led U.S. companies and report cost increases in the United States of 18 to 115 percent. Nine executives reporting cost increases in the United States manage non-U.S. companies; the range of their increases was 25 to 85 percent. Both the top and the bottom of the range for non-U.S. companies were reported by Canadian executives in the same industry—paper and other forest products.

In a few cases, executives report overall cost increases for more than one of their company's products. A British consumer products executive, for example, reports that one product increased considerably more than a second for all the countries cited, including the United Kingdom (Chart 4). In contrast, a Canadian paper company executive reports that the increases for two of his company's products made in Canada and the United States were almost the reverse of each other. The production costs for product one went up 80 percent in Canada over a five-year

Table 1: Five-year Production Cost Increases—Percentage (1971 is base year)*

Companies Countries	United States															Canada								
	A	B	C	D	E	F	G	H	I	J	K	L	M	N	O	A_1	A_2	B	C	D	E	F	G	H	I_1	I_2
United States	72	51	35-40	50	51		74	46	35	115	50	18-20	55	61	65-75	80	25									
Canada			75		50		62	31		90				61		70	67	50	99	121	72		73	80	90-100	90	55	85	60
Japan													48													
Belgium														213												
France				90	104									157							43					
Germany					98	38								112											55	
Italy				125	71									200							125					
Netherlands	60							65			50	21		149												
Sweden				90	92																					
Switzerland																										
United Kingdom		75		80	79			74				134		138							128					135

Table 1: Five-year Production Cost Increases—Percentage (1971 is base year)* (Continued)

Countries \ Companies	Belgium A	Belgium B	Belgium C	France A	France B	Germany A	Germany B	Germany C	Netherlands A	Netherlands A¹	Sweden B	Sweden C	Switzerland A	Switzerland B	Switzerland C₁	Switzerland C₂	Switzerland C₃	UK A₁	UK A₂	UK B	UK C	UK D₁	UK D₂	UK E₁²	UK E₂²	UK F³
United States	40																	59								49
Canada										45										52						
Japan																					61					
Belgium	79	139	58																					74	113	
France	108	98	135	80								80												112	137	
Germany			45	28	53			27		25											73			102	115	
Italy			158								115	100				37	60				38			73	139	
Netherlands						38																				
Sweden										36	80	80				58	39							109	153	
Switzerland	167												38	15–20	46									120	217	
United Kingdom																		63	84	130	103	130	103	72	115	113

¹ 1970 is base.
² 1969 is base—6 years.
³ 1972 is base.

*Subnumbers indicate different product lines.

Source: The Conference Board.

Table 2: Five-year Production Cost Increases—Percentage Range[1]

Country	Total Number of Instances	Range of Increases %	Number of Home Country Instances	Range of Increases %	Number of Other Country Instances	Range of Increases %
United States	23	18-115	14	18-115	9	25-85
Canada	18	31-121	11	50-121	7	31-90
Japan	2	48-61	0		2	48-61
Belgium	6	58-213	3	58-139	3	74-213
France	11	43-157	1	80-135	10	43-157
Germany	12	25-115	3	27-53	9	25-115
Italy	11	37-200	0	-	11	37-200
Netherlands	6	21-149	1	38	5	21-149
Sweden	7	36-109	3	36-80	4	90-153
Switzerland	7	15-217	5	15-58	2	120-217
United Kingdom	18	63-167	9	63-130	9	74-167
	121	15-217	50	15-139	71	21-217

[1] 14 instances reported by one company cover 6-year period.
Source: The Conference Board, 1976-1977.

Chart 4: Six-year Production Cost Increases—a U.K. Consumer Products Company

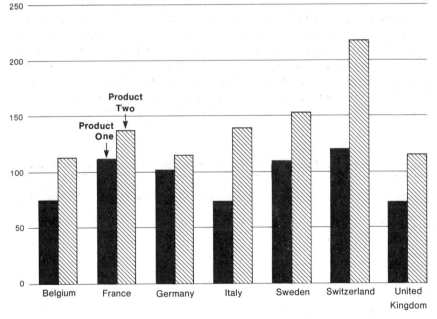

Source: The Conference Board, 1977

Table 3: Five-year Production Cost Increases—Rubber Industry

Country	U.S. Company A	U.S. Company B.	UK Company	Swedish Company
United States	50	51	52	—
Canada	—	50	—	—
Japan	—	—	61	—
France	90	104	73	—
Germany	—	98	38	—
Italy	125	71	—	115
Sweden	90	92	—	80
United Kingdom	80	74	103	—

Source: The Conference Board, 1976-1977.

period and 55 percent in the United States. The costs for product two went up 60 percent in Canada while rising 85 percent in the United States.

Variety and broad range also characterize the responses within particular industries. In the rubber industry (reading across Table 3) two American and one British executives report similar cost increases in the United States, but considerable variation in the United Kingdom and in France. One of the American executives reports the highest increase of the three companies in France and the lowest in the United Kingdom, while the British executive reports the lowest increase of the three companies in France and the highest of the three in its home country. (These comparative figures must always be viewed with caution, several executives pointed out, because while the products made in each country and the product mix may be similar, they are almost never identical.)

The two U.S. rubber companies and a Swedish company produce in

Table 4: Five-year Production Cost Increases—Chemical Industry

Country	U.S. Company	Belgian Company	Swiss A Company	Swiss B Company
United States	115	—	—	—
Canada	90	—	—	—
Belgium	—	58	—	—
France	—	98	—	—
Germany	—	45	—	—
Italy	—	158	—	37-60
Switzerland	—	—	15-20	39-58

Source: The Conference Board, 1976-1977.

both Italy and Sweden. One of the two American executives reports a substantially lower cost increase in Italy than either the Swedish or the other American executive. At the same time, the same American executive reports the highest increase of the three in Sweden although the range of differences there is much smaller than in Italy.

Figures reported by four chemical executives for their overall cost increases are shown in Table 4. Again, considerable differences are seen. One Swiss executive reports cost increases in his home country more than twice as high as those of a second Swiss executive. The first executive also reports his company's cost increases in Switzerland to be very close to those it had in Italy. In contrast, a Belgian company shows cost rises in Belgium greatly below those it experienced in Italy, where its increases were more than two-and-a-half times those of its Swiss competitor.

One of the factors of immediate concern to executives who must decide where to produce a particular product is the cost of manufacturing a product at any one time in one country compared with that for the same product in another country—assuming access to the same market is available in the option. Several executives participating in this study make such cost comparisons for manufacturing a few products outside their home countries in relation to the cost for manufacturing them at home. A detailed compilation of the data they submitted is given in Table 5, and a summary is reported in Table 6.

Of the 76 facilities reported on, 19 are located in the home countries of the reporting executives and their costs are used as the index base. Of the remaining 57 units located outside the home countries, 32 are reported to have higher production costs than the home country unit, 23 have lower costs, and production costs in two instances are the same as in the home country.

A look at individual country reports shows considerable differences. Executives from the United States report that 17 of their 20 foreign manufacturing units had production costs higher than those at home for the same products. In the production unit in Italy, costs were lower for only one industry; in two industries, with facilities in Canada and the United Kingdom, costs were the same as in the United States. Executives in the United Kingdom and Sweden similarly report production costs more often higher in their foreign units than in their home facilities. In contrast, Canadian and Swiss executives report production costs abroad were usually lower than those at home. In the case of Canada, all seven units (six in the United States and one in the United Kingdom) were reported to have lower costs than the Canadian manufacturing facilities at home for the same products. And in the case of Switzerland,

Table 5: Overall Cost Relationships—1976 (Home Country=100)*

Country \ Company	United States				Canada						Belgium	Sweden			Switzerland		United Kingdom		
	A_1	A_2	B	C^1	A	B_1	B_2	B_3	C_1	C_2	A^1	A_1	A_2	B^1	A^1	B	A_1	A_2	A_3
United States	100	100	100	100	80	90	80	70	65	85					67		118	90	101
Canada	118	100	118	103	100	100	100	100	100	100		160	156		78				
Japan	252	133													90		101	77	104
Belgium			107	191							100				98				
France	128	117		144							83			140					
Germany		136		155							91	98	126		86	91	131	126	155
Italy	91	102		123							103			140	86				
Netherlands				112											104				
Sweden												100	100	100	86				
Switzerland															100	100			
United Kingdom	130	101		100						80					81		100	100	100

[1]1975
*Subnumbers indicate different product lines.
Source: The Conference Board.

Table 6: Summary of Overall Cost Relationships, 1975 or 1976

Home Country of Company	Total Number of Units Compared	Number Located in Home Country	Number Located in Other Countries	Other Country Costs Compared with Home Country:		
				Higher	Lower	Same
United States	24	4	20	17	1	2
Canada	13	6	7	—	7	—
Belgium	4	1	3	1	2	—
Sweden	11	3	8	6	2	—
Switzerland	12	2	10	1	9	—
United Kingdom	12	3	9	7	2	—
	76	19	57	32	23	2

Source: The Conference Board.

nine of the ten units abroad had lower costs than their home production units; only in the Netherlands did it cost a Swiss company more to produce a product than in Switzerland and the difference was small.

3.
Components of Rising Production Costs

RAW MATERIAL COSTS, labor costs—including wages, salaries and benefits and energy costs are the three components of rising production costs at home most frequently mentioned by executives. For some executives only one or two factors are commanding. A Canadian machinery manufacturer states: "As for the major causes of the increase, by far the most significant was the cost of materials. Second in order of significance is labor cost. Other cost components are generally insignificant." And an executive from a Belgian steel products firm observes: "In order of importance, cost increases are mainly due to personnel costs (wages and salaries), raw material cost, and energy cost."

Other factors contributing to rising production costs at home, but mentioned far fewer times, include lower productivity, new government regulations, insufficient capacity utilization, and the costs of packing materials, transportation, and subcontracting.

The order of significance of the three leading contributors to increasing production costs varies from company to company. For those executives who listed a priority of any kind, raw material costs or labor costs led the list with one or the other taking first place. Energy costs most commonly took third place.

Summarizing the comments from executives from all ten countries, a rough geographical pattern appears. Among Canadian executives raw material costs more frequently head the list of factors contributing to high costs at home, while among European executives increasing labor costs are more often cited in first place. Among U.S. executives, the priority listing is more equally divided between the two main components.

The same components, and often in the same listed order, contribute to increasing production costs for their foreign units, according to most corporate officers who identified components contributing to increasing total production costs. A vice president of a U.S. rubber company, discussing the rising costs in several European countries, writes: "The principal causes for these increases in cost correspond to those factors affecting our domestic production cost in the same order of significance, (i.e., raw materials, labor and energy)." The chairman of a competing British rubber company lists "raw materials, employee costs, energy costs" in that order as the reasons for his company's increases in the United Kingdom. He lists the same reasons in the same order for production units in the United States and in Japan and the same reasons—but with raw materials and employee costs changing priority positions—for production in Germany and France.

Raw Material Costs

The 1970's have seen a substantial increase in the cost of raw materials used in almost every industry in almost every country. Table 7 shows the five-year increases in raw material costs reported by several participants in this study. The range reported is from 10 percent for one product line manufactured in the United States by a U.S. company to 186 percent reported by a Belgian executive for his company's production unit in the United Kingdom. The industries represented in the table are optical goods, forest products, rubber products, industrial machinery and equipment, food and beverages, automobiles, other steel products, and glass.

Figures reported by the OECD at the end of 1976 for selected European countries bear out the company reports for this study. The OECD shows heavy increases in raw materials prices from 1970 to 1975 (Chart 5). Belgium and France experienced declines from the year before in such prices in 1975, but the prices for raw materials in that year were still substantially above 1970. In both Italy and the United Kingdom, prices climbed steadily throughout the period after 1970 and most rapidly after 1973 to the point where executives in Italy faced raw material prices in 1975 almost three times those of 1970.

A number of executives observe that knowing the increased costs of raw materials (or any other component by itself) is insufficient to determine the effect of that factor on overall production costs. Raw material costs may have increased greatly, but may be a relatively small part of the total cost of making a particular product. The increase in the raw material costs, consequently, may be relatively insignificant for the total production cost of the product. The financial vice president of a U.S. optical company illustrates such a situation: "In ophthalmic glass lenses,

Table 7: Five-year Increases in Material Cost—Percentage*

Country \ Company	United States									Canada						Belgium	France		Germany	United Kingdom
	A_1	A_2	A_3	B	C	D	E	F		A_1	A_2	B	C	D_1	D_2	A	A_1	A_2	A	A^1
United States	30-35	10	48	53	23-47	79	64	82						45	95	49				41
Canada				42		63				68	99-117	110	90	90	70					
Belgium																57				
France				78				128								100	122	135		
Germany				65			69												23	
Italy				57																
Sweden				77																
United Kingdom				76			138	64								186				127

[1] 4 years.
*Subnumbers mean different product lines.
Source: The Conference Board.

Chart 5: Raw Material Prices

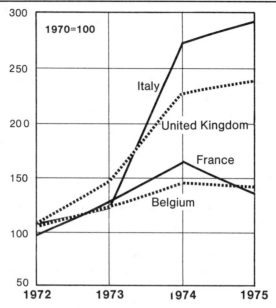

Source: Organisation for Economic Co-operation and Development, Main Economic Indicators, December, 1976

materials increased 48 percent while direct labor increased 25 percent; however, in this product line, labor and overhead comprise a larger proportion of total cost than does materials. Therefore, the labor increase is more significant in terms of total production costs."

A financial executive of a Canadian machinery manufacturer cites the opposite situation—one in which raw material costs are a large portion of total costs so that any increase in raw materials costs has a major effect on total costs: "Because material comprises 80 to 90 percent of our total product cost structure, this invariably is the major cause of increase, whether or not it has increased more than other cost components in percentage terms."

The chief executive officer of a second Canadian company (forest products) submits figures illustrating "the relative importance of the cost increases after they have been weighted according to their percentage of the total cost." With lumber, for example, the five-year percentage increases place the raw materials component (105 percent) lower than wages and salaries (115 percent). When these components are weighted for their importance to the total cost per unit produced, the increased cost of raw materials (55 percent) becomes higher than that for wages and salaries (40 percent). The same reversal occurs in the cost of producing pulp. The five-year actual increase is 80 percent for raw materials and 115 percent for wages and salaries, but the weighted increases are 30 percent for raw materials and 25 percent for wages and salaries.

Labor Costs

Sharing responsibility with rising material costs for a significant portion of the total increase in production costs in the 1970's is the rapid increase in labor costs including wages, salaries and benefits. The rise in labor costs has been reported by international organizations such as the OECD and the European Community. Table 8 shows the rise in unit labor costs in manufacturing for selected developed countries, as reported by the OECD; and Table 9 shows the increase in manufacturing labor costs, as reported by the European Community, using a common accounting unit to make its comparison among six countries.

Wage increases are a major contributor to total production cost rises for a large majority of companies, and this is true across a wide range of industries. The chairman of a Canadian industrial products company explains: "Average hourly rated wages in our company increased by 63.1 percent from the beginning of 1972 to the end of 1976. During the same period, the Consumer Price Index rose by 42.1 percent, giving our employees a substantial gain in wages." The president of a Canadian metals company points out that "labor costs are currently 65 percent of [our] Canadian production costs," and adds: "Total labor costs, including benefits, have increased from $6.00 per hour to $11.30, or 88 percent, over the period."

Many corporate officers from the United States and Europe, as well as Canada, include labor costs prominently on their lists of components of total increased production costs. An executive of a European manufacturer of medical instruments observes: "Production costs have increased by 53 percent. Since our production is labor intensive, the

Table 8: Index of Unit Labor Costs in Manufacturing (1970=100)

	1972	1973	1974	1975
United States	101	104	114	127
Canada	104	109	121	137
Japan	118	122	157	192
Sweden	124	139	163	204
United Kingdom	113	123	147	197

Source: Organization for Economic Cooperation and Development, Main Economic Indicators, December 1976.

Table 9: Manufacturing Labor Costs—1972-1975 (in Eur—a common accounting unit)

	1972	1973	1974	1975
Belgium	320	370	450	550
France	269	310	340	420
Germany	348	410	470	520
Italy	270	280	310	n.a.
Netherlands	344	400	490	570
United Kingdom	n.a.	208	240	270

n.a.—not available

Source: European Economic Community, Luxembourg, Eurostat, *Statistical Telegram,* 1/1976.

main causes were sharp and steady increases of wages (53 percent) and of social welfare costs due partly to government regulations, partly to agreements between employers' organizations and trade unions (e.g., social insurance for old age, disability, sickness and unemployment; paid leave for vacation and sickness; vacation and Christmas bonuses; and others). Social welfare costs increased from 53 percent of wages in 1971 to 62 percent in 1976. Costs of raw materials and energy have increased at lower percentages."

The five-year wage increases experienced by these companies are shown in Table 10. Among the 20 instances reported by U.S. managers, the range of labor cost increases is from 17 percent for a manufacturing unit in the United States to 131 percent for a U.S. company's unit in France. For the 12 instances cited by Canadian executives, the range is from 41 to 140 percent—with both the top and the bottom of the range being for units in Canada itself. For the 37 instances covered by European executives, the range of increase is from 32 percent for a British company's U.S. unit to 200 percent for a French company's unit in Italy. Table 11 summarizes the ranges of these five-year wage increases by country.

Several executives note that the effects of wage increases on total production costs sometimes can be offset, at least in part, by increases in labor productivity or other productivity factors. An executive of a U.S. consumer products company asserts: "It should be noted that wage increases in the direct labor areas have been substantially offset by productivity improvements gained by capital investment in our high-volume, three-plant U.S. manufacturing network." Two European executives report the offset is more limited for their companies. The president of a rubber company reports: "Wages increased in 1972 by 12 percent and in 1973 by 6 percent. The effect, however, was reduced by improvements in productivity and changes in product mix." And the

Table 10: Five-year Wage Increases—Percentage*

Country \ Company	United States								Canada									Belgium	France		Germany		Italy	United Kingdom					Netherlands
	A_1	A_2	A_3	B	C	D	E	F	A	B_1	B_2	C	D	E	F	G_1	G_2	A	A_1	A_2	A	B	A	A_1	A_2	B_1	B_2	C^1	A
United States	17–24	20	25	27	49	30	57	51										35			70			32					53
Canada				49		54			88	41	72	110	63	140	80	45	70						50					50	
Japan																120	60												
Belgium				131			105											120	125		130		100						127
France				81			63	93										126	100	150	100	115	90						113
Germany				104															70	75		52	50						88
Italy																			200				120						185
Netherlands																													98
Sweden				59																									102
United Kingdom				73			108	88									130	144					100	116	75	108	103	112	156

[1] 4 years.
*Subnumbers indicate different product lines.
Source: The Conference Board.

Table 11: Percentage Range, Five-year Wage Increases by Country

	Number of Instances	Range—Percentage
United States	15	17- 70
Canada	12	41-140
Japan	1	130
Belgium	5	100-127
France	8	90-150
Germany	9	50- 93
Italy	4	104-200
Netherlands	1	98
Sweden	2	59-102
United Kingdom	12	73-156

Source: The Conference Board.

head of a European textile company states: "There has been some modest increase in output per employee over the period which has slightly reduced the impact of wage increases."

Managers of other companies report, however, that they have experienced little or no gain in productivity to offset wage increases. Labor costs per ton of paper for one product line went up 140 percent, according to an executive of a Canadian forest products company. He adds: "Productivity in tons per man-hour has remained unchanged during this period."

The Bureau of Labor Statistics of the U.S. Department of Labor has

Table 12: 1975 Indexes of Manufacturing Output and Unit Labor Costs (1967=100)

	Manufacturing Output	Unit Labor Costs in National Currency	Unit Labor Costs in U.S. Dollars
United States	106.3	156.4	156.4
Canada	136.1	151.1	160.3
Japan	168.3	222.8	272.0
Belgium	154.7	150.8	204.2
France	136.7	195.0	224.1
Germany	137.1	164.7	267.5
Italy	139.4	254.2	243.2
Netherlands	138.5	175.0	249.8
Sweden	137.1	167.6	208.9
Switzerland	118.3	149.2	150.2
United Kingdom	112.9	247.1	199.6

Source: U.S. Department of Labor, Bureau of Labor Statistics, Office of Productivity and Technology, January, 1977. (Later published in the *International Economic Report of the President,* 1977.)

reported on the increase in unit labor costs (both in national currencies and in U.S. dollars) since 1967, and the increase in manufacturing output for the same years (Table 12). Only Belgium is reported as having a lower increase in unit labor costs than its increase in output, and this is reported only for labor cost increases in national currency; when measured in U.S. dollars, the unit labor cost increase exceeds the increase in manufacturing output.

Executives from a few responding companies provide data comparing labor costs for their foreign production units with those in their home country units. These data are shown in Table 13. Except for the German labor costs for one product line of a U.S. company and the notable exception of the Dutch company, labor costs for this small number of foreign units are generally reported lower than those at home. Executives of two U.S. companies report labor costs for their product lines in Canada lower than labor costs for the same products at their home facilities in the United States. At the same time, a Canadian executive reports that labor costs for two of his company's product lines in the United States are lower than the labor costs for the same products at the company's home facilities in Canada.

Table 13: Labor Costs Compared, 1975 or 1976 Using Home Country as 100*

Company Country	U.S. A 1976	U.S. B 1975	Canada A_1 1976	Canada A_2 1976	Belgium A 1975	Italy A 1975	Netherlands A 1976
United States	100	100	70	75	—	—	121
Canada	90	94	100	100	—	—	—
Belgium	—	69	—	—	100	—	114
France	—	77	—	—	59	—	78
Germany	105	81	—	—	75	96	117
Italy	—	59	—	—	60	100	71
Netherlands	—	—	—	—	—	—	100
Sweden	—	—	—	—	—	—	137
United Kingdom	64	56	—	—	—	63	56

*Subnumbers indicate different product lines.
Source: The Conference Board.

Energy Costs

The rising cost of energy is the third major component of increasing production costs. The effect of rising energy costs depends to a large extent on the industry and the amount of energy it uses. On the one hand, a European executive of a firm making electrical goods observes: "The energy crisis in 1974 was roughly responsible for an extra increase of 7 percent. This was due to indirect effects (inflation and material costs) because in most of our products energy costs are very low (1 to 2 percent)." On the other hand, the president of a Canadian basic metals company reports that all fuels and power account for 15 percent of present production costs. He goes on to state that over a five-year period hydropower costs increased 77.8 percent, fuel oil 119.6 percent, natural gas 132.7 percent, and metallurgical coke 178.7 percent.

In the cement industry, a European executive states that two-thirds of the five-year increase in manufacturing costs at his company is due to price increases for fuel (oil, gas) and energy. A British textile executive notes that energy costs for his company had increased twice as much as wages (270 percent to 135 percent) in the years between 1971 and 1976, and much more than raw material costs (150 percent).

The vice president of a U.S. glass manufacturing company raises another point of concern regarding energy costs. He notes that increases in labor and raw material costs have had the biggest impact on total production costs "but energy has created the largest concern in the manufacturing process. Not only is the cost of energy a factor, but the fluctuating availability has created the need to invest in the capability of having alternate energy sources available at every production location."

A few executives report that their companies have been able to adjust to increased energy costs due to circumstances that may be peculiar to their companies and countries. A Swiss executive notes: "Energy costs have remained relatively stable due to long-range supply contracts, as well as internal energy production in hydropower plants." And a German explains: "Without the use of nuclear energy it would not have been possible to limit the rise in fuel costs for the period 1970 to 1976 to about 50 percent. Long-term supply contracts for natural gas, under which the price adjustment has not followed the increased energy price yet, contributed to this limited rise in costs as well."

Other Components of Rising Production Costs

A few of the executives mention cost factors other than the main three to which they attribute partial responsibility for rising production costs. A European producer of energy cites *new environmental regulations* and *insufficient capacity utilization* as two of these factors. On the first, he notes: "Additional costs in crude oil processing have resulted from higher standards—compared to the rest of the world—of environmental protection set for the operation of refineries and quality of products. Since 1970, electricity generation costs have increased by approximately

60 percent, owing to the rise in fuel costs on the one hand, but also due to the considerable increase in capital expenditure, which is not least attributable to the charges resulting from environmental protection requirements." Commenting further on the rising costs of processing crude oil, he states: "Costs per ton processed increased as a result of insufficient capacity utilization of the processing plants in the past two years. Capacity utilization of the refineries decreased far below 70 percent because of increased imports of petroleum products."

The president and managing director of a Canadian basic metals producer cites increasing *transportation costs* for their effect on production costs: "The cost of transportation for ores and concentrates per unit of final product has increased due to fixed plant locations and the mix of ores from the various mines. Total transportation costs have increased 86.5 percent in the period." He points out that these rising transportation costs constituted 5.5 percent of total product costs.

Presidents of consumer products companies in the Netherlands and the United Kingdom attribute part of their rising product costs to the increasing costs of *packaging materials* which, in turn, are affected by increasing labor, materials and energy costs.

Components of Rising Production Costs Outside the Home Country

In explaining their experience with increasing costs abroad, executives cite the three factors at work in their own countries—increasing costs for labor, raw materials, and energy. An executive of a U.S. consumer products company recounts his firm's experience in several countries.

In the Netherlands: "Cost increases experienced in 1975 again result from material costs dependent on energy costs, combined with relatively small volume gains in facility utilization resulting in inflated unit overhead costs."

In Japan: "Production unit costs expressed in local currency have increased less than 8 percent per year, with the exception of 1974 in which costs increased 38 percent due primarily to the impact of energy costs on material costs."

In England: "Cost increases result from material costs dependent on energy costs, as well as substantial wage increases from strong local inflation. While sizable productivity improvements have been made, they have not, however, been able to keep pace with wage increases."

And, finally, in Mexico: "Production unit costs expressed in local currency have increased 83 percent since 1971, with a 32 percent increase experienced in 1973 attributable to energy cost impact on materials cost. Secondly, higher overheads and wages from local inflation have not been offset by proportionate productivity improvements."

Slower growth, no growth, or sometimes decline of productivity are cited by a number of executives as a factor in the rise of production costs. One U.S. executive reports the five-year change in productivity in his firm compared with its rising labor costs and total factory costs. His report for seven countries is shown in Chart 6. In three European plants output increased, but at a far slower pace than labor and total costs. In Canada and two European countries output actually declined while costs rose. At home, there was no gain in productivity in the face of a 27 percent increase in direct labor costs and 51 percent increase in total factory costs.

An industrialist in Milan compares his company's productivity and labor cost in Italy with his experience in Germany and the United Kingdom (Chart 7). In both Germany and the United Kingdom the cost of labor is lower than it is in Italy and productivity is higher. He elaborates: "I would like to point out that the gap that exists in productivity is due

Chart 6: Five-year Percentage Increases—Costs and Output (U.S. company—1971 is base year)

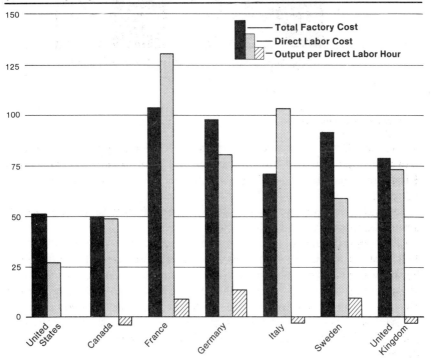

Source: The Conference Board

Chart 7: Comparison of Productivity and Labor Cost in 1975 (Italian company)

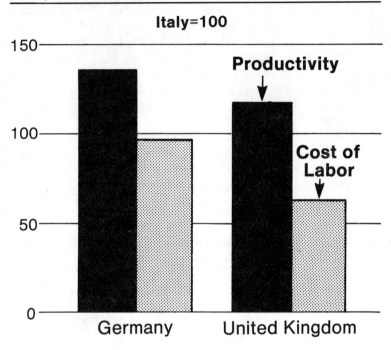

Source: The Conference Board

to the fact that in Germany operatives work an average of 25 to 30 days more per year and one hour more per day than in Italy; in the United Kingdom they work about 20 days more per year and half an hour more per day."

Several executives note other factors contributing to the differences among countries in production costs.

• Exchange differences, for example, are cited by a vice president of a U.S. optical instruments company as accounting partly for differences between Canada and the United States: "Since the Canadian dollar also strengthened during this period, the relative costs of production between the two countries has favored the United States over the five years."

• And an American metal products executive blames absenteeism for adding to higher labor costs and, therefore, to higher production costs in the Netherlands.

4.
Effects of Higher Production Costs

ALTHOUGH the public may expect major companies to take drastic action in the face of rapidly rising production costs, there has been little or no change in company operation patterns, according to many executives. "Basic changes in the structure of our company or the shifting of production locations have neither been carried out nor are they intended," states a German executive. An American executive writes: "These changes in costs really do not have that great an effect on where we locate facilities or where we make our products."

Reasons for No Changes

A Canadian executive attributes no production changes following increasing production costs to the fact that such costs are going up all over. He explains: "Trading patterns have not been affected to any great extent so far by the recent increase in production costs as a result of inflation because the economies of most western countries in which the company has operations have also experienced high costs from inflation."

Rising costs are important, but other factors are equally, and sometimes more, important in determining production and trading patterns and policies. A U.S. executive points out: "Decisions relating to optimum sourcing of product must consider many factors in addition to basic production costs (import restrictions, duties, export subsidies, chauvinistic purchasing, and so on). At this time, the greater majority of our production is in the United States and we have no present intention to reduce U.S. sourcing."

A Canadian beverage products executive must recognize acquired tastes in countries where his firm operates. Because of these tastes, the company cannot bring in lower cost beverages manufactured in some other country. He explains: "Relative cost changes among our various production points have had little to do with our company's patterns of production and trade. The reason for this is that our products, although similar in nature, are *not* the same; they are not substitutes for one another."

Typically, chief executives see the decision on where to manufacture more in terms of marketing requirements than in terms of cost. "We do not expect any changes in the placement of manufacturing facilities due to differences in production costs," observes the executive vice president of a U.S. metal products firm, adding: "The placement of our productive capacity is primarily a marketing decision." And the chairman of a European rubber company asserts: "Changes in patterns of production and trade have been and will be largely dictated by market requirements and competitors' activity."

More detailed explanations by three executives for their decision not to alter existing manufacturing and trading patterns despite rising costs are found in the box below.

Executive Reports on Reasons for not Altering Manufacturing Locations Despite Rising Production Costs

A Belgian Chemical Executive

In general the manufacturing cost increases of our various products in the different European countries have not had—and likely will not have in a near future—a big impact on the choice of the location of our plants and their development, or on our trading policy.

Our business is essentially based on commodities (soda ash, chlorine, caustic soda), or intermediates, and on commodity plastics. These products are manufactured in large-size, capital-intensive plants. Therefore, it is a heavy type of activity, difficult to move—except maybe over a long period—from one place to another to benefit from more favorable local and short-term conditions.

On the other side, the use of important quantities of heavy raw materials makes it necessary to place the manufacturing units near the supply sources. This is another obstacle to the plants' mobility.

Social factors also have an influence in the same way. In Europe, several legal and contractual rules do not allow us to put workers out of a job. Their transfer from one place to another, especially through different countries, is a most difficult problem to solve.

Financial conditions also have to be taken into account. In many European countries investments receive state aid if made under stated conditions. Stopping or moving this kind of activity to other business centers is generally opposed by the government.

A Swiss Chemical Executive

At least equally important as a factor in competition as production cost increases due to inflation are changes in currency exchange rates. Since 1971, the nominal external value of the Swiss franc against all other currencies has increased sharply—against the dollar by some 60 percent. Even in terms of purchasing power (i.e., taking account of inflation rates in Switzerland and abroad) the external value of the Swiss franc has increased appreciably relative to all market regions. This means that Switzerland has become relatively less favorable as a production site. It is, therefore, possible that expansion of capacity will be more likely to take place abroad than in Switzerland. However, investment cannot be governed by short-term developments. At present Switzerland has the lowest inflation rate of any country. This means that the internal upward pressure on prices is lower than elsewhere, with a consequent rise—provided that the external value of our currency does not increase further—in the international price competitiveness of Swiss exports. There are, moreover, only a few countries in which new investments are not called into question by inflation, nationalization and other government intervention. Relocation of existing production from Switzerland abroad is therefore scarcely attractive.

A Canadian Machinery Manufacturing Executive

Because of very high investment requirements for new manufacturing facilities and in order to obtain the benefits of large-scale production, the company has tended to rationalize production of similar products into relatively few manufacturing sources around the world, using these established locations as a base for expansion. Of course, we are influenced by governmental legislation in many developing countries requiring a greater or lesser degree of local manufacture. Hence, cost of production has not been the significant factor in determining location of manufacturing facilities. Additionally, inflation rate differentials between countries tend to be offset in the long term by relative changes in the value of currencies—offsetting any short-term benefits.

The determination of trade patterns between these source manufacturing locations and our markets around the world has been an evolutionary process. Because of past product differences from different manufacturing sources we have been loath to change this pattern as it undoubtedly causes problems in the market place.

Changes Resulting from Higher Production Costs

Several executives report they have made some changes as a result of recently increased and increasing production costs. The changes cited include relocating manufacturing, adopting new technologies, locating new raw material sources, tightening management, changing products, and, in one instance, divesting.

Relocating Manufacturing

Cost increases and the changing relationships of production costs among nations have led a number of European executives to turn to the United States to set up new manufacturing facilities or to expand already existing ones there. A Dutch executive comments: "The United States has had the lowest cost increase. The devaluation has made it practically impossible to export from the EEC to the United States except for certain very specialized products. So we increased our local production activities in the United States."

Two Swiss executives also turned to the United States. One reports: "Switzerland has undoubtedly become a high-cost country, which increases the tendency to manufacture parts of the production in other areas. The cost disadvantage is mainly felt in the markets belonging to the dollar area. With respect to this situation we are investigating possibilities to enlarge our local manufacturing in the United States." And the chairman of the second Swiss firm, which had a new plant under construction in the United States at the time of writing, explains: "Production in Switzerland is further handicapped by the high value of the

Swiss franc. This makes our products more expensive in the export markets and attracts lower priced competition in the Swiss home market. As almost all our raw materials are imported, the effect of the high Swiss franc is basically limited to the added value. As a result, new production capacities will be built abroad rather than in Switzerland."

A German executive reports moving production facilities to a developing country and shifting other manufacturing around in Europe as a result of rising costs and changing cost relationships. He observes: "Rising wages at our traditional production facilities and increasing competition from underdeveloped countries at low price levels have forced us to capitalize on all rationalization opportunities and to look for cheaper production locations. In order to stay competitive we established production facilities in Malaysia in 1972 and moved one part of our production line from Spain to Malaysia, another part from Germany to Spain."

An American executive, too, reports shifting some of his firm's manufacturing around Europe for cost reasons. He notes: "In Europe, where trade barriers are being reduced and geographical proximity enable us to ship economically, we are regionalizing purchasing and reassigning manufacturing activities between plants to take advantage of lower wage rates in our more labor-intensive products."

Technology Changes

Several executives report that their companies are making technological changes in manufacturing to cut costs. A Dutch executive notes his firm is making such changes, leading to higher output per hour. An executive from the United States explains his company was changing over to automated types of equipment to reduce labor input.

Two Canadian forest product executives report taking similar technological steps to improve their production cost positions. The president of the first explains: "In Canada, new technology in thermomechanical pulping has resulted in the installation of improved refining equipment in two of our newsprint mills in order to reduce the cost of raw materials. Capital will be spent in Canada on modernizing our pulp and paper mills and building materials conversion mills in order to reduce costs, improve value added, and improve our competitive position." And an executive of the second firm observes: "We have attempted to improve productivity by means of increased mechanization and automation and to lower manufacturing costs through the development and use of new techniques such as thermomechanical pulping."

Related to the adoption of new technology is the decision reported by two European steel products executives to shift their firm's production more heavily into high-technology products in order to maintain their cost competitiveness. An executive describes one of the measures adopted by his company as the "creation and development of more sophisticated products to be manufactured in the European plants." A second steel products executive reports: "In the processing sectors we are going to concentrate our efforts—also with respect to our investments—to a higher degree on highly sophisticated products."

Other Changes

A number of other changes in production operations resulting from higher production costs in the last few years are mentioned by several executives. A Canadian reports that his company is looking for new and cheaper sources of raw materials and finding them "in lower cost areas such as Brazil and the Far East."

Two other Canadian executives report efforts by their companies as a result of cost pressures to make current and new production facilities more efficient. One of them points out his company's steps "to move to more integrated operations, e.g., lumber mills closer to pulp and paper operations and fiber sources, finishing facilities closer to or integrated with our fine paper manufacturing operations." Still another Canadian company reported steps taken to introduce "far greater standardization to facilitate sourcing changes."

An Italian executive reports his company has diversified its products to mitigate the effects of rising production costs: "The changes that have taken place in production costs require a continuous effort in order to control competitiveness of products and manufacturing processes. In this context our group has concentrated its efforts on diversification in the satisfaction of the demand for social needs (e.g., ecology, housing, public transport), and in higher technology products."

Finally, one Canadian president reports the disposal of two of his firm's European plants—one in the United Kingdom in 1976 and one in Italy "some years ago." Costs were a factor in these divestments. He explains: "Rising costs, coupled with weak market conditions in 1975 and 1976, have resulted in a strong squeeze on profit margins." And in 1976, the company experienced an operating loss for the first time in its history.

5.
Production Costs in the Future

MOST EXECUTIVES expect production costs for their firms both at home and abroad to continue to increase in the years immediately ahead. At the same time, several report, they expect increases for the next five years to be more moderate than those of the last five. A German executive comments: "On the whole, the cost development in the sectors under review will be below the rates of increase of the past five years. This is mainly attributable to the fact that the crude oil price is not expected to rise as drastically as in 1973-1974. With a slowdown in economic development and competitive pressure also, the remaining costs will not increase to the extent experienced in the past." Also taking this position is the president of a Canadian company: "We expect production costs to continue to increase in most of our domestic and foreign operations over the next five years, but not as dramatically as during the past five years." Similarly, an executive of a Danish firm states his belief that the increases of production costs in the coming five years will be slightly less than in the past five years.

Few executives are ready to give specific figures for production costs projected five years ahead. The forecasts of those that did are shown in Table 14. In addition, a few executives report their expected average annual increases for production costs over the next five years. Three U.S. executives report 7 to 8 percent, 8 to 9 percent, and 7 to 10 percent cost increase per year for their U.S. operations. A Canadian executive reports 3 to 4 percent for his firm's Canadian units, and a German executive reports 5 to 10 percent for his company's German facilities. The vice president of one U.S. company gives no estimate for his firm's projected cost increases in the United States, but reports its European plants, with

Table 14: Estimated Percentage Increases in Overall Production Costs over the Next Five Years

Country \ Company	United States			Canada	Europe			
	A	B	C	A[1]	A[1]	B	C[1]	D
United States	50	50	35	38	—	—	—	27
Japan	—	—	35	—	—	—	—	14
Belgium	—	—	—	—	47	—	—	—
France	—	—	—	—	57	—	[a]30-60	—
Germany	—	—	—	—	30	20	—	37
Italy	—	—	—	—	81	—	30-60	5
Netherlands	—	—	35	—	—	—	—	—
Sweden	—	—	—	—	—	—	30-60	—
United Kingdom	—	—	70	—	—	—	—	55

[1] 4 years
[a] Varies by product line.
Source: The Conference Board.

increases from 80 to 125 percent in the past five years, "may experience cost increases of about two-thirds the rate for the past five years."

The reasons given for the anticipated continued growth of production costs in the five years ahead are the same reasons used to explain increases in the five years just ended—primarily increases in the costs of labor, raw materials, and energy. A Canadian executive, for example, writes: "We do expect that the principal causes of the cost increases that we will experience in the near future will be the same as those for the recent increases, i.e., raw materials—specifically wood—wages and energy."

A German executive explains: "Further cost increases have to be expected in the sectors mentioned in the next five years." They will largely depend on the development of the energy level on which the individual consuming countries have hardly any influence. Although costs are expected to be more stable in the Federal Republic than in other countries, capital expenditures will increase about 4 percent per annum. Higher requirements to be met with regard to the safety and environmental protection of power plants will probably result in rates of increase clearly above that rate. Wage and salary costs per employee are expected to increase some 7 percent per annum in the next five years, compared with 9 percent per annum between 1971 and 1976."

Expected Labor Cost Increases

Several executives, in addition to the German above, cite anticipated increases in labor costs. A Canadian expects increases to average 7 percent a year in Canada. A second Canadian anticipates increases of 6 to 8 percent per year in Canada and adds: "In the United States we would not expect costs to rise faster than in Canada, and they may increase at a slightly slower rate due to a more stable labor environment. In the United Kingdom, a forecast of the increase in wages and salaries is made difficult by the existence of the 'social contract.' The Government has committed itself to increases of 6 to 8 percent per year, but it is questionable whether this can be attained."

A British executive sees a decline in the rate of labor cost increases in the United Kingdom in the years immediately ahead. He reports expected increases for labor costs of his company in the next few years as 12 percent in 1977, 12.8 in 1978, 8.7 in 1979, 7.1 in 1980, and 6.4 in 1981.

A Belgian executive reports anticipated average annual labor cost increases for several countries in which his firm operates. In Belgium he expects the average annual increase to be 11 to 12 percent, 13 percent in France, 10 in the United Kingdom, 10 in Japan, and 11 in the United States. Using 1969 as a base, a British executive projects his company's anticipated labor costs for several countries for the years 1976 through 1979. These data are shown in Table 15.

Because labor costs will increase at varying rates, the relationships of these costs among countries in which executives have producing units will continue to be important. One executive, from a company in the Netherlands, presents such relationships for a future year. His index figures, with the Netherlands as the base, are shown in Chart 8, along with the labor cost relationships among the same countries for 1976. Labor costs in the United States, Belgium, Germany and Sweden exceed those in the Netherlands in 1976 and are expected to exceed them in 1980. By

Table 15: Estimated Labor Costs per Unit of Output—a British Company (1969=100)

	1976	1977	1978	1979
Belgium	409	442	509	606
France	376	451	515	596
Germany	208	238	278	321
Italy	278	278	345	443
Sweden	406	465	562	686
Switzerland	317	362	424	513
United Kingdom	232	278	322	368

Source: The Conference Board.

Chart 8: 1976 and 1980 Wage Cost Relationships for a Dutch Company

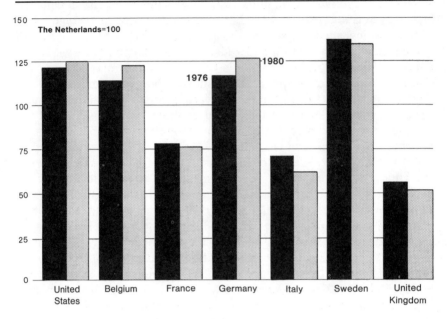

Source: The Conference Board

the latter year the difference, he expects, will increase for the United States, Belgium and Germany, while it will decrease slightly for Sweden. He notes, however, that the Swedish costs for his company still are the highest when compared with the home country. In 1976, the costs in France, Italy and the United Kingdom are all below those in the Netherlands. By 1980, they are expected to be even lower when compared with the home country's costs.

Expected Raw Material and Energy Cost Increases

A few executives made specific forecasts of expected increases in raw material costs over the next five years. A Canadian estimates the increases will average 6 percent a year in Canada. A British executive sees the increase rate declining from 15 percent in 1977 to 8 percent in 1978, 7 in 1979, 6 in 1980, and 5 in 1981, for his company in the United Kingdom. And a Belgian executive anticipates an annual increase of 7 to 8 percent in Belgium, 8 percent in France, 5 in the United Kingdom, 6 in Japan, and 7 in the United States.

The same British and Belgian executives also report their estimates of expected cost increases for energy during the next five years. The British officer expects increases for his facilities in the United Kingdom to be 15 percent in 1977, 10 in 1978, 9 in 1979, 8 in 1980, and 7 in 1981. The Belgian reports anticipated annual increases in energy costs for the next five years will average 8 percent in Belgium, France and Japan; 10 percent in the United Kingdom; and 12 percent in the United States.

Possible Changes Resulting from Future Production Cost Increases

In the expectation that production costs are going to climb in the years immediately ahead, a number of executives indicate possible reactions by their firms to increased production costs. These range from more weighty consideration of production costs in the selection of *new* manufacturing locations to no changes at all for current trading and manufacturing patterns and policies.

Effects on New Manufacturing Facilities

Several executives report their firms will be giving more attention to production costs in the next five years and that such costs will play an increasingly important, if not determining, role in selecting locations for new or expanded manufacturing facilities. A U.S. executive writes: "During the next five years we intend to evaluate critically the establishment of new overseas facilities based on the production cost level, plus other important considerations that are prevalent during the specific period that we will be considering new overseas ventures."

And a Swedish executive states: "The location of manufacturing facilities in different countries has so far been governed by the demands from our customers for local manufacture. However, manufacturing cost considerations will become more important in the future with the present high wages in Sweden."

Some of the executives indicate that rising costs are among the factors leading their firms to locate new manufacturing units to serve specific markets. A Dutch executive, after discussing the components of rising production costs, concludes: "Their probable effect will be that we are going to produce more and more in the market areas where we are selling." The president of a Canadian company similarly states: "We certainly feel that rapid inflation of production cost factors—and its uneven pattern—is a great obstacle to the expansion of international trade. Great instability in exchange rates is also a serious impediment. The result is that we shall have to be much more careful than in the past regarding the location of basic production facilities, and will also tend more than in the past to develop facilities within each important market. All this may well result in a lower rate of growth for the future."

A number of executives from Canada believe the United States is a more probable place for the future growth of their production facilities than their homeland. This is especially true for the forest products industry. An executive of one such firm reports: "New investment in plant and equipment will most likely occur in the United States unless significant cost savings can be obtained in Canada by fully integrating existing 'nonintegrated' operations. We expect that the United States will still be the low-cost producers with respect to Canada over the next five years."

The president of a second Canadian firm in this industry highlights the effects on production costs, and consequently on plant location, of rising labor costs: "In the medium term, it is evident that increases in Canadian labor rates must be held to levels substantially lower than those which occur in the United States. Unless labor can be made to recognize the existing dangers, this objective may be difficult to achieve without a continuation of the current labor dislocations. Faced with this situation, many Canadian firms are presently directing a greater proportion of their discretionary capital expenditures budget towards opportunities that exist in the U.S. market."

An executive of a U.S. machine parts company agrees with the Canadians on the current attractiveness of investing in the United States. His firm has producing units in Ireland and on the European continent. He observes: "I anticipate in the future that the costs in Europe will be very similar to those in the United States and, in Ireland's case, possibly will exceed U.S. costs. New manufacturing facilities will be established only in the United States."

Another American executive also believes the future production costs of his firm, as well as other factors at home and abroad, will favor production in the United States. This may not mean new production facilities, he reports, but may mean increased exports: "I believe that relative inflation rates as well as foreign exchange currency rates will continue to strengthen the position of U.S. exports versus local manufacturing in most European countries."

Other Effects

A U.S. executive, whose company has been automating more and more of its production in order to reduce labor input, states that his firm will continue to automate in the hope of reducing still further growing labor costs. A Canadian executive, who also suggests: "judicious use of capital for automation," reports another tactic to be pursued by his company: "A major part of maintaining our competitiveness includes product redesign to use less expensive material, less labor, and less expensive manufacturing processes. This probably will affect our pricing strategy more over the next five years than conventional cost reduction of existing products."

No Changes

Policies and plans for the location of new manufacturing plants and existing trade patterns will not be changed for their companies, according to several executives, even with continuing increases in production costs. "We do not foresee," a Swedish executive notes, "any important changes in the placement of our manufacturing facilities or in trade patterns during the next five years, even though we do foresee further increases in production costs both at home and in foreign countries due to increased cost for labor, raw materials, and energy." The vice president of a U.S. company concludes: "The trading patterns of our company are not anticipated to change significantly over the next five years. We will continue to supply the United States demand from domestic facilities and the European demand from our Italian facilities."

Factors other than production costs play a major role in the determination of where future production facilities will be built or expanded. At times, these other factors outweigh the cost factors. This is the main reason, according to several executives, for reporting no change in their firm's international production plans despite the increase in production costs at varying rates at different production sites.

A senior British businessman cites several factors, other than production costs, that enter into his firm's decisions on the placement of new production units. He explains: "No specific answer can be given as to the effects of these expected cost trends on the location of new manufacturing facilities and pattern of trade, but the following are important in this connection:

(1) The decline in U.K. cost competitiveness should be offset by exchange rate movements, at least in the short term. For this reason we are planning for increased exports from U.K. sources for many products.

(2) Our U.K. divisions are the centers of worldwide businesses which depend on the U.K. for technical guidance, and must be in the forefront of technology and efficiency. Therefore, they have a major claim on investment funds.

(3) The transportation costs and marketing characteristics of many of our products are conducive to a pattern of operations in which each major market is supplied by its own local manufacturer. This obviously limits the scope for planning manufacture and sourcing on an international scale."